Daniel Adamczyk

Calculation of the Cosmic Expansion Rate

GRIN Publishing

Bibliographic information published by the German National Library:

The German National Library lists this publication in the National Bibliography;
detailed bibliographic data are available on the Internet at http://dnb.dnb.de .

Imprint:

Copyright © 2012 GRIN Verlag GmbH
Print and binding: Books on Demand GmbH, Norderstedt Germany
ISBN: 978-3-656-86298-7

This book at GRIN:

http://www.grin.com/en/e-book/285449/calculation-of-the-cosmic-expansion-rate

Foreword

My idea for the accelerated expanding universe can not exist without the fourth spatial dimension of the Kaluza-Klein theory, not without a lot of strained balloon model and not without the power of a Big Bang explosion. One idea for the dark matter is a major component of it, because without it no accelerated expansion is possible. It would have made no sense to me to design a cosmological principle if I nad not found a way for the natural mass rejection and with which the Dark Energy at first is to explain.

Crucial for the invention of this cosmological model is the speed of the enlargement of the balloon, which represents the dark matter as a relative increase in mass. The speed of movement, the velocity in fourth dimension, is by the character of the room in which we live, invisible. Unfortunately we will never be able to verify this view, since we are now living in the balloon skin.

So the model is admittedly risky, because it requires many things which can never be put in appearances. Only circumstantial evidence can corroborate this view. It remembers to science of the structure of elementary particles. But even the summoning of this effort is not enough. In recognition of the Kaluza-Klein theory it would be necessary to put themselves in the situation of the inhabitants of this universe, to capture their mental perspective, and to prove so the view of this article.

That's asking a lot, but the calculation reproduces the actual observation of the expansion rate very good. Probably we will never be in a better position to observe a universe. So I think a study of this view would be useful and necessary.

Remarks

The natural mass rejection theoretical is only to make possible by a mixture of the theories of EINSTEIN and NEWTON of gravitation

The r_n/r-Term means a stereographic projection of a four dimensional ball on a three dimensional room – it is just an attempt!

Results

In this calculation the universe is the skin of a four-dimensional ballon. All the mass is gathered in the infinitesimally thin skin. But this skin is a three-dimensional room of which every point has the same distance to the centre of the balloon.
With this view one can see a gravitation on the distant skin of the centre because for calculation one can unit the whole mass in the centre of the ballon. Under this thoughts one can find a distance R from the centre where the time ist still standing to a place far far away from the centre; that is a place without gravitation. This distance R is an Event Horizont of a Black Hole, which has no rotation and is not electric charged. This R is calculated by $R = M \cdot G/c^2$. G means the gravitational constant, c the speed of light, but M means in this understanding not the rest mass but the rest mass plus the Dark Matter. The Dark Matter is gravitational effective, so it effects on time by gravitational time dilation.
If we do anticipate the results of this calculation, one can see the distance of the universe to the centre of the balloon r to the Event Horizont R, so it looks like the following graphic:

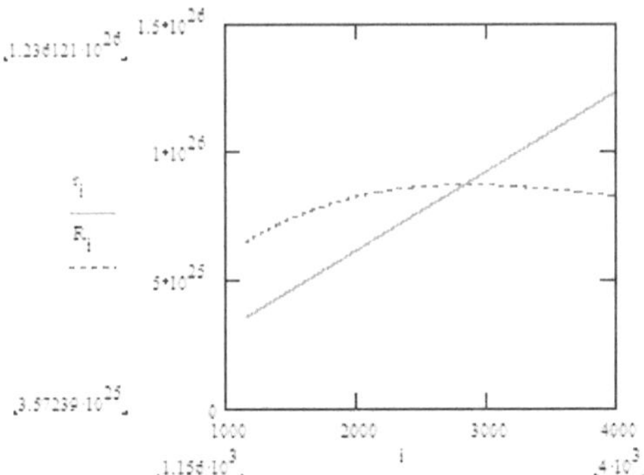

The index i this means only the steps of the calculation. The steps i refer to the distance r from the center of the balloon and not to the age of the universe. But step 4000 is our present, so one can see that the distance of the skin to centre is bigger than the distance of the Event Horizont. This means, the universe is not a Black Hole at this time, but one can see it has been one in the past where r < R. The calculation shows that this is at least 3.8 billion years ago.

There where two solutions to calculate our distance r of the present, but it was easy to find out, which is the right one to us. Science found the average densitiy of baryonic matter , ρ_{bary} , in the cosm. In my calculation Dark Matter results of the speed, matter has over the enlargement of the balloon. One can consider, that this speed has to be near the speed of light if you watch the factor q between baryonic matter and baryonic plus Dark Matter, because it is calculated by

$$M = \frac{m}{\sqrt{1 - v^2/c^2}} \text{ so that } v = c \cdot \sqrt{1 - m^2/M^2}$$

Here m means the rest mass and v is the velocity of the enlargement of the balloon. The factor q results of the measurement of the densities of barionic and Dark Matter, which where

$$\rho_{bary} \approx 4.5 \cdot 10^{-28} \, kg \cdot m^{-3} \qquad \text{and} \qquad \rho_{bary+dark} \approx 3.0 \cdot 10^{-27} \, kg \cdot m^{-3} .$$

The next step to understand the calculation of rest mass of the universe is a little harder to reconstruct. But for the fist view one can consider the following. If a balloon inflates with an inflation speed v and someone is standing on the balloon and watches a point on the balloon, he will watch a escape velocity of the point which is the bigger as more far away the point is. The speed of escape v_e is calculated by $v_e = \varphi \cdot v$. φ is the angle of centre of the balloon between the point of view and the viewed point. However, this presupposes that the inhabitant of the skin can watch round the curve of the skin, but this is something that will have a big significance later. But at first a graphic for explanation:

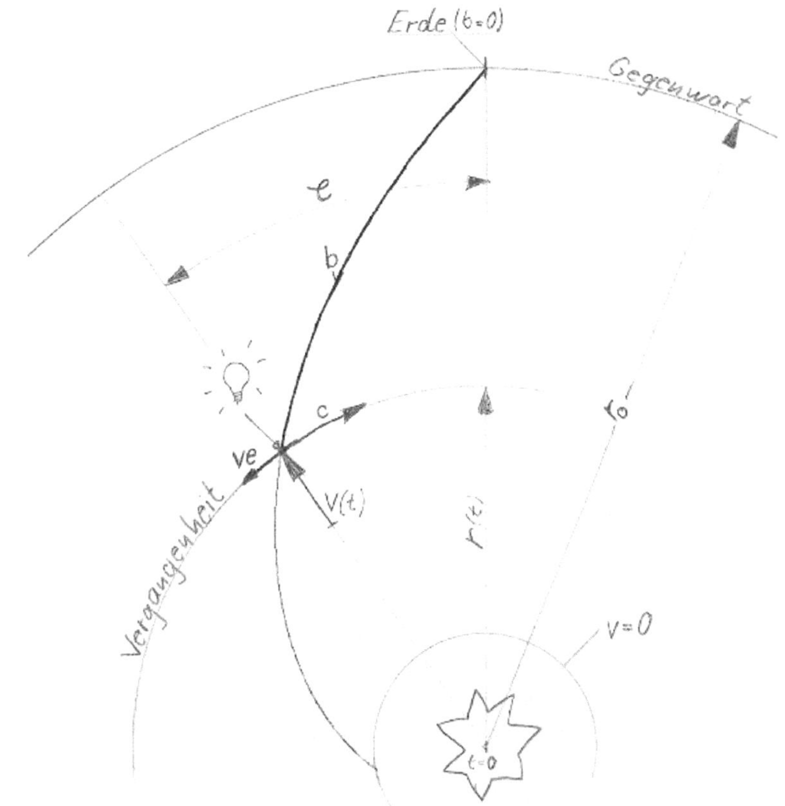

Now the speed of escape v_e shall be the basis of the cosmic expansion rate H, which is measured by science. But to define $H = v_e/s$ there is additional the distance to the object neccesary. If one consider, that the light of the far away object wants to reach the observer while the skin inflates, you understand by the picture of an ant, that tries to reach another point of the inflating ballon, that a) she always has the same velocity against the skin and b) the time to reach the point takes the same time as the difference between the start- (r_u) and end-size (r_n) of the inflating balloon. The way of the distance s bases on the same time as the way of the enlargement of the inflating balloon so that one can say for a constant v and c

$$s = c \cdot t \text{ and } r_n - r_u = v \cdot t \text{ results in } s = \frac{c}{v} \cdot (r_n - r_u)$$

Nevertheless, without φ a calculation of H is not possible. φ is a result of a mathematical problem called logarithmic spiral on which can to fall back because the way of the light is under an angle to the skin of the inflating balloon.
If you consider that light tries to reach the observer while the ballon inflates, one can see that the path does not only happen in three dimensions of the cosm but also in the fourth dimension of the balloon. So the way s lays oblique in the room of the balloon and also to the

cosm, the skin. By the way, this is no problem to imagine because of we know that we look through the past if we watch distant objects – the skin of the balloon stands for the present, the present which we never can see because we look to the past and which is only one point in the world of an individual - but every individual has one! But there is to say that the visible universe geometrical is not a sphere – it is a flame, or a drop.

It is easy to understand, that the angle α of the way towards the direction of movement of the skin is $\alpha = arc\tan(c/v)$. Now, in a logarithmic spiral the angle φ is calculated by

$$\varphi = \frac{1}{k} \cdot \ln\left(\frac{r_u}{r_n}\right).$$ The k represents the angle α by $k = \cot(\alpha)$ so $\varphi = -\frac{c}{v} \cdot \ln\left(\frac{r_u}{r_n}\right),$

if one consider, that the spiral shall wind from the present to the past, as it is necessary for the calculation of this essay. So k has to be negative. r_u is the distance of the remote point to the center of the balloon, as this is seen by observers in the past and u means a step of i.

If one remembers the reason of this excursion he will understand that it is now to estimate the rest mass. This succeeds by verifying the distance of the skin to the centre. It is a complicated calculation which needs the arc length of the logarithmic spiral, but not enough, one has to derive from the arc length to the way of light s mathematically, not logical as we had done it. I offer as a way out my experience of our equation of s. Many attempts confirms that in the equation of s the start distance has to set $r_u = 0$ that you get $r_n = s \cdot v/c$ as the distance, where is the rest mass is hidden in the Hubble-Constant or cosmic expansion rate.

With an expansion rate $H_0 = 74km \cdot s^{-1} \cdot MPc^{-1}$ the distance r of the present skin is $r_n = 1.236 \cdot 10^{26} m$ because one knows the relationship between baryonic plus Dark Matter and baryonic matter and therefore v . This leads over a volume of a three-dimensional sphere $V = 2 \cdot \pi^2 \cdot r^3$ to a rest mass of $m = 1.681 \cdot 10^{52} kg$. Additional there is to say that the deviation to the full calculation is low because of the big explosion power of the Big Bang which accelerates the mass very strong.

In science it is a fact that we look directly to the centre of the cosm. So we think that Hubble-Constant represents the rayon of the universe. By the type of this connection the science is confirmed. So it is a nice feature. In a different form it is shown by

$$r_n = \frac{c}{H_0} \cdot \sqrt{1 - \frac{\rho_{bary}^{\,2}}{\rho_{bary+dark}^{\,2}}} \quad \text{and if one call} \quad q = \frac{\rho_{bary+dark}}{\rho_{bary}} \quad \text{it is} \quad r_n = \frac{c}{H_0} \cdot \sqrt{1 - 1/q^2}$$

in which H_0, ρ_{bary} and $\rho_{bary+dark}$ has to refer only on really observed measurement of the local region in memory of the single point of presence which has to be fullfilled. The upper panel allows the question of whether the light can shine even in the past - the answer is, the light can only shine in the three dimensional sphere, our cosm. It is transported just by the inflation of the ballon, which always expands to the future. The only thing which is able to reverses is the size of Dark Matter – what follows additional.

To close this chapter the cosmic escape velocity v_e is shown down over the distance s as it is calculated:

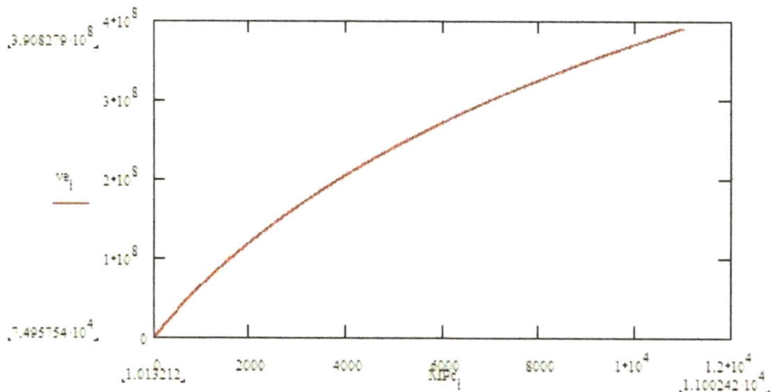

For comparision the results of measurement is shown down:

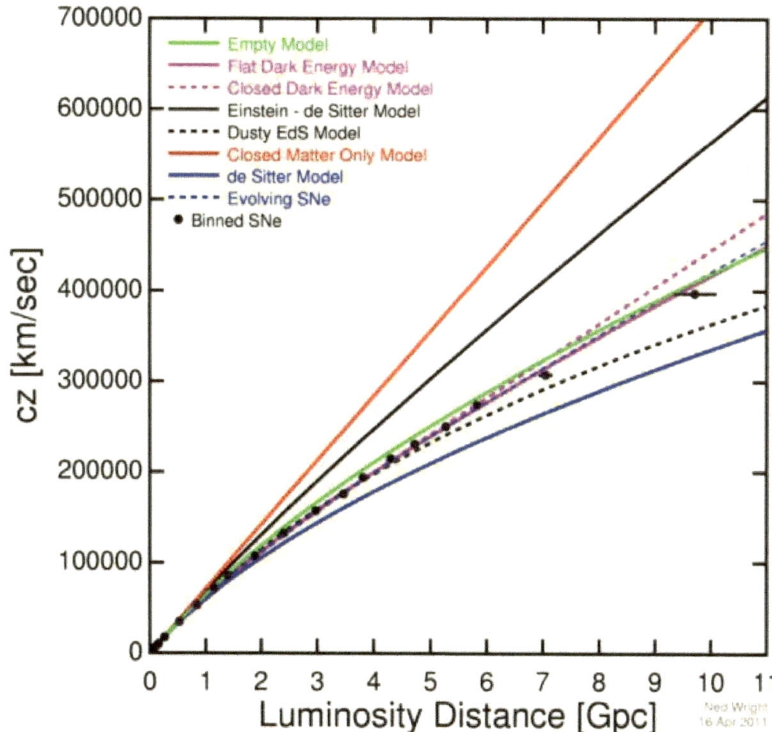

Natural mass rejection

Before having a look to the calculation of the expanding rate we have to make an excursion through the gravitational acceleration, which is already done two times, namely by EINSTEIN and NEWTON. But this view needs the two existing theories. So that it is already based on these two sublimed thougts.

At first it is necessary to hear the view of engineer Walter Orlov who calculates the relativistic mercury perihelion with 50 arcseconds against EINSTEINS 43". Later it will be shown that the calculation with the equation of natural mass rejection compensates for the difference of the seven arcseconds, so that Orlov has no reason to doubt EINSTEINS General Theory of Relativity.

"For the calculation of planetary orbits, we the Schwarzschild Metric used. At large distances from the heavy masses they are easier to Minkowski metric. Why is always meant that the effects of special relativity theory in general Relativity theory are considered automatically. in the general picture, it may vote yes. But if we look at a specific bill more closely, we know that there are additional assumption is needed to discover. Although this damage already mentioned public. Thus, for the solution of the relativistic Kepler problem, the proper time used - I.e. the time in the reference system of the planet: "Then the orbital parameters s is the time τ , indicating an entrained Clock " Prof. Norbert Dragon [26] - so that the Lorentz factors for Energy and angular momentum are removed:

$$d\tau = dt \cdot \sqrt{1 - \frac{v^2}{c^2}} \Rightarrow \frac{dt}{d\tau} = \frac{1}{\sqrt{1 - \frac{v^2}{c^2}}} .$$

On this way you get

$$E = \frac{m \cdot c^2}{\sqrt{1 - \frac{v^2}{c^2}}} \Rightarrow E = m \cdot c^2 \cdot \frac{dt}{d\tau} ,$$

and of

$$L = \frac{m \cdot [\vec{r} \times \vec{v}]}{\sqrt{1 - \frac{v^2}{c^2}}} \quad \text{results} \quad L = m \cdot c^2 \cdot [\vec{r} \times \vec{v}] \cdot \frac{dt}{d\tau} = m \cdot r^2 \cdot \frac{d\varphi}{dt} \cdot \frac{dt}{d\tau} = m \cdot r^2 \cdot \frac{d\varphi}{d\tau} .$$

The effects of special relativity theory are thus from the bill is simply removed and left a simple energy equation without Lorentz factors:

$$\frac{m \cdot r^2}{2} + \frac{L^2}{2 \cdot m \cdot r^2} - G \cdot \frac{M \cdot m}{r} - G \cdot \frac{L^2 \cdot M}{m \cdot c^2 \cdot r^3} = \frac{E^2}{2 \cdot m \cdot c^2} - \frac{m \cdot c^2}{2}$$

They contend that the omission of the factors specifically Relativity theory has no effect on outcome. The good at so it is a metric that the change of the reference systems the quality of the equations does not change. It is expected therefore, the same result for example in the reference system of center of gravity. Let us introduce the bill is not there because they have a times would have been too complicated. All that is true, however. Anomalous perihelion of the planet by the observations determined for the reference system of gravity. Also finds the solution of the Kepler problem for the movement principle of celestial bodies in the reference system of gravity place. Two-body problem is the two single-body-problems converted: 1. Movement of center of gravity passes as a whole no acceleration, therefore, is simply ignored; second (Actual Bill) movement of a body with reduced mass $\mu = m_1 \cdot m_2 / (m_1 + m_2)$ in the outer potential $U = G \cdot (m_1 + m_2)/r$, that moves along with the main focus.

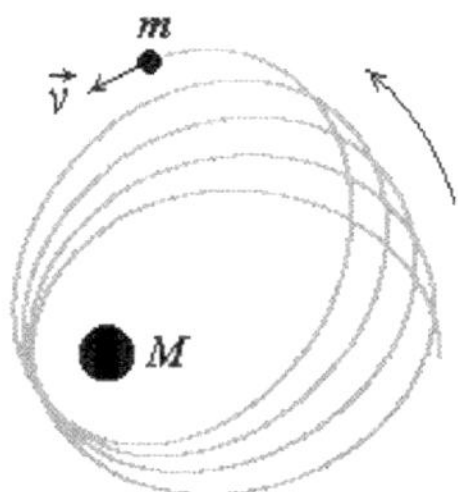

Furthermore, it is not the consideration of relativistic mass as difficult as it is meant. We assume that the classical energy equation with r^{-3}-term, according to general relativity theory Mercury, for instance, already 43 "returns. Relatively is the mercury-perihelion for turning around very small. Therefore we can only calculate only the addition, by r^{-3}-term leave out, but relativistic energy and angular momentum use (to later add to this addition to 43 "):

$$E = \frac{m \cdot c^2}{\sqrt{1 - \dfrac{v^2}{c^2}}} - G \cdot \frac{M \cdot m}{r} \ ,$$

With the solution of the equation is concerned e.g. Prof. Schnizer [27] and in principle we do the same now. The Lorentz factor we can express it by the angular momentum and then in the Energy use rate:

$$L = \frac{m \cdot r^2 \cdot \dot\varphi}{\sqrt{1 - \dfrac{v^2}{c^2}}} \Rightarrow \dot\varphi = \frac{L}{m \cdot r^2} \cdot \sqrt{1 - \frac{v^2}{c^2}} \ , \qquad\qquad \dot r = \frac{dr}{d\varphi}\dot\varphi = r' \cdot \dot\varphi \ ,$$

$$\frac{v^2}{c^2} = \frac{\dot r^2 + r^2 \dot\varphi^2}{c^2} = \frac{\left(r'^2 + r^2\right) \cdot \dot\varphi^2}{c^2} = \left(\frac{L}{m \cdot c}\right)^2 \cdot \left(\frac{r'^2}{r^4} + \frac{1}{r^2}\right)\cdot\left(1 - \frac{v^2}{c^2}\right) .$$

With the new variable $s = \dfrac{1}{r}$, $s' = -\dfrac{r'}{r^2}$

we get

$$\frac{v^2}{c^2} = \left(\frac{L}{m \cdot c}\right)^2 \cdot \left(s'^2 + s^2\right)\cdot\left(1 - \frac{v^2}{c^2}\right)$$

and at least

$$\frac{1}{\sqrt{1 - \dfrac{v^2}{c^2}}} = \sqrt{1 + \left(\frac{L}{m \cdot c}\right)^2 \cdot \left(s'^2 + s^2\right)} \ .$$

This term we use now in the energy equation:

$$E = \frac{m_0 c^2}{\sqrt{1 - \frac{v^2}{c^2}}} - G\frac{Mm_0}{r} = m_0 c^2 \sqrt{1 + \left(\frac{L}{m_0 c}\right)^2 (s'^2 + s^2)} - GMm_0 s \, .$$

It follows

$$(E + GMm_0 s)^2 = m_0^2 c^4 \left[1 + \left(\frac{L}{m_0 c}\right)^2 (s'^2 + s^2) \right]$$

$$E^2 + 2EGMm_0 s + (GMm_0)^2 s^2 = m_0^2 c^4 + L^2 c^2 s'^2 + L^2 c^2 s^2$$

$$s'^2 = \left(\frac{E}{Lc}\right)^2 - \left(\frac{m_0 c}{L}\right)^2 + \frac{2EGMm_0}{L^2 c^2} s - s^2 \left[1 - \left(\frac{GMm_0}{Lc}\right)^2 \right] \, .$$

Now we use

$$\lambda^2 = \left[1 - \left(\frac{GMm_0}{Lc}\right)^2 \right]$$

and it follows

$$\left(\frac{ds}{d\phi}\right)^2 = \left(\frac{E}{Lc}\right)^2 - \left(\frac{m_0 c}{L}\right)^2 + \frac{2EGMm_0}{L^2 c^2} s - s^2 \lambda^2$$

$$\lambda d\phi = \frac{ds}{\sqrt{\left(\frac{E}{Lc\lambda}\right)^2 - \left(\frac{m_0 c}{L\lambda}\right)^2 + \frac{2EGMm_0}{L^2 c^2 \lambda^2} s - s^2}}$$

The Swing from r_{min} to r_{max} and back supplies of the right side 2π. It follows

$$\lambda(\phi - \phi_0) = 2\pi, \quad \phi - \phi_0 = \frac{2\pi}{\lambda}.$$

Thus, the perihelion shift in the direction of rotation (figure below) is equal:

$$\Delta\phi = \frac{2\pi}{\lambda} - 2\pi = 2\pi(\lambda^{-1} - 1) =$$

$$= 2\pi\left[\frac{1}{\sqrt{1 - \left(\frac{GMm_0}{Lc}\right)^2}} - 1\right] \approx \pi\left(\frac{GMm_0}{Lc}\right)^2$$

This is per cycle $\Delta\phi \approx 8.35 \cdot 10^{-8}\, rad$ and per century about 7".

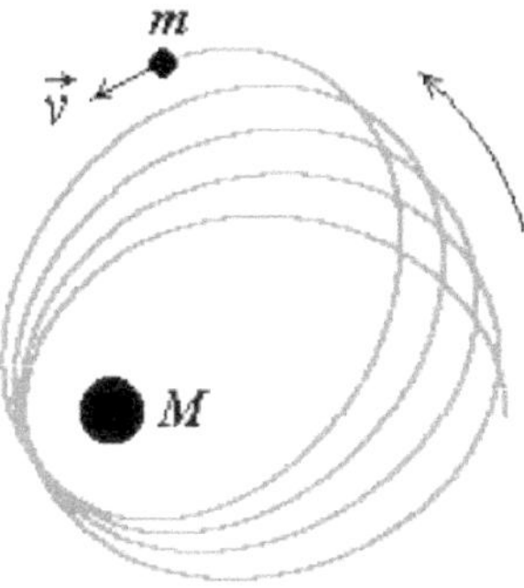

Figure. The perihelion shift in the direction of rotation.

In this way according to relativity theory complete calculation of the perihelion precession of Mercury takes 50" instead of 43" in the century.

Now it is useful to have a look on my own derivation of gravitation, which includes the natural mass rejection. To this end we consider the free fall:

First, we note that potential energy is $E_{pot} = m \cdot U$. In this chapter m is the mass of the specimen and U means the gravitational potential. After this we remember NEWTONS law of the free fall $E_{pot} + E_{kin} = const.$ Therein we set v = 0 so that is $E_{kin} = 0$. Then one can say $const = E_{pot(R)}$. Here R means the starting height of the fall where v = 0. It seems not very intelligent to tell now that $E_{kin(r)} = const - E_{pot(r)}$, but I want to show by a look over $E_{kin(r)} = E_{pot(R)} - E_{pot(r)}$, that $E_{kin(r)} = \Delta E_{pot}$. With the knowing that r is the current altitude while falling, and remembering $const = E_{pot(R)}$, is it comprehensible way, that $\Delta E_{pot} = E_{pot(R)} - E_{pot(r)}$.

To prevent future issues is to say that the start value must be subtracted from the target value.

The following thoughts shows the idea on which all further considerations are based. EINSTEIN told us mass is equal to energy and left us the equation $E = m \cdot c^2$. His general theory of relativity states that, although the speed of light is constant, but only locally at each site. Furthermore, it varies as follows: $c' = c \cdot \left(1 - \Delta U / c^2\right)$.

Now it is to calculate a difference of the mass equivalent: $E' = m \cdot c'^2$ and further $\Delta E = E - E'$, so that one can say

$$\Delta E = m \cdot c^2 - m \cdot \left(c \cdot \left(1 - \frac{\Delta U}{c^2} \right) \right)^2$$

The problem with doing this is, that both energies E and E' are on different heights. It is to say, that EINSTEIN tells us, that one and the same specimen has different energies in different heights. In my opinion the difference ΔE creates the gravitational field plus gravitational acceleration or motion in free fall. 'Matter must always have the same rest mass energy.'

Using this example of free fall, this was $\Delta E = E_{start} - E_{target}$. Showed with some theoretical attempts to

$$\Delta E_{pot} = \frac{1}{2} \Delta E .$$

Now it is to develop a potential out of this to drawn conclusions to the gravitational acceleration. Like it is done with the Newtonian derivation of gravitation the starting height is infinitely. So the gravitational potential is zero. So if you call the mass of the planet M

$$\Delta U = \frac{-M \cdot G}{r_{start}} - \frac{-M \cdot G}{r_{target}}$$ you can call it $\Delta U = \frac{M \cdot G}{r_{target}}$. Thus

$$\Delta E = m \cdot c^2 - m \cdot \left(c \cdot \left(1 - \frac{M \cdot G}{r_{target} \cdot c^2} \right) \right)^2$$ and with $\Delta E_{pot} = \frac{1}{2} \Delta E$ it is logical to say

$$\Delta E_{pot} = \frac{1}{2} \cdot \left(m \cdot c^2 - m \cdot \left(c \cdot \left(1 - \frac{M \cdot G}{r_{target} \cdot c^2} \right) \right)^2 \right)$$

If we know that $E_{pot} = m \cdot U$ one can say that $\Delta E_{pot} = m \cdot \Delta U$, which means that

$$m \cdot \left(U_{start} - U_{t\,arg\,et} \right) = \frac{1}{2} \cdot \left(m \cdot c^2 - m \cdot \left(c \cdot \left(1 - \frac{M \cdot G}{r_{t\,arg\,et} \cdot c^2} \right) \right)^2 \right) .$$

Regarding the derivation of NEWTON $U_{start} = 0$. That means

$$m \cdot \left(- U_{t\,arg\,et} \right) = \frac{1}{2} \cdot \left(m \cdot c^2 - m \cdot \left(c \cdot \left(1 - \frac{M \cdot G}{r_{t\,arg\,et} \cdot c^2} \right) \right)^2 \right)$$

and

$$- U_{t\,arg\,et} = \frac{1}{2} \cdot \left(c^2 - \left(c \cdot \left(1 - \frac{M \cdot G}{r_{t\,arg\,et} \cdot c^2} \right) \right)^2 \right)$$

so it is easily to say that

$$U_{t\,arg\,et} = - \frac{1}{2} \cdot \left(c^2 - \left(c \cdot \left(1 - \frac{M \cdot G}{r_{t\,arg\,et} \cdot c^2} \right) \right)^2 \right)$$

Now the newtonian gravitational potential is done by

$$U = \int_{-\infty}^{r} g(r)\, dr$$

So it is easily to say, that the gravitational acceleration g is

$$g = \frac{d}{dr} U_{target} .$$

Thereupon

$$g = \frac{d}{dr} - \frac{1}{2} \cdot \left(c^2 - \left(c \cdot \left(1 - \frac{M \cdot G}{r_{t\,arg\,et} \cdot c^2} \right) \right)^2 \right)$$

and finally

$$g_{target} = \frac{M \cdot G}{r^2} - \frac{M^2 \cdot G^2}{r^3 \cdot c^2}$$

The derivation includes the view, that the rest mass energy communicates over differences in height. A specimen shall know his constant sum of its energies and forms part of which the gravitational field and forms part of which a force or a motion.

Without experimental evidence this statement remains ineffective. But actual is to say that this equation of gravitational acceleration can assume negative values. Next will be shown a theoretical attempt with the mercury perihelion by usage of the equation of g above. The calculation is based on the principle of free fall according to the NEWTONian view. The loop is shown down:

$$\varphi := \begin{array}{|l} vz \leftarrow 0 \\[4pt] r \leftarrow 46 \cdot 10^{9} \\[4pt] vu \leftarrow 5.9 \cdot 10^{4} \\[4pt] \varphi \leftarrow 0 \\[4pt] \text{while } \varphi < 2 \cdot \pi \cdot d \\ \quad \begin{array}{|l} a \leftarrow \dfrac{M \cdot G}{r^{2}} - \dfrac{M^{2} \cdot G^{2}}{r^{3} \cdot c^{2}} - \dfrac{vu^{2}}{r} \\[8pt] vz \leftarrow vz + a \cdot t \\[4pt] rv \leftarrow r \\[4pt] r \leftarrow r - vz \cdot t \\[4pt] vu \leftarrow vu \cdot \dfrac{rv}{r} \\[8pt] \varphi \leftarrow \varphi + \dfrac{vu \cdot t}{r} \\[8pt] \text{break if } ((\varphi > 2 \cdot \pi \cdot (d - 0.1)) \cdot (r > rv)) \end{array} \end{array}$$

The velocity vz is the speed of mercury in the direction to the sun, vu is the rotation velocity of mercury. φ is the back down angle of mercuy round the sun. The program starts with $\varphi = 0$ in the perihel and rounds the sun for nearly hundred years namely 415 rounds. The program uses more accurate values as shown above. The compiler of this calculation is Virtual Pascal v2.1 and the program is shown in the attachment. The screen below shows the results of the program steps of t = 0.1 seconds:

XXXXXXXXXXXXXXXXXXXXXXXXXXXXXXXXXXXXXXXXXXXXXXXXXXXXXXXXXXXX
XXXXXXXXXXXXXXXXXXXXXXXXXXXXX

 Noch 0 Minuten

Aphel
Abstand M-S = 4.66763975543473E-0001 AE
Geschwindigkeit= 3.88582621045506E+0004
Umlaufzeit= 8.79690137409616E+0001 Tage
Umlaufanzahl= 4.14499994484433E+0002   f= 4.14500000000000E+0002
Anzahl Durchgaenge= 3.15041665338111E+0010
Apheldrehung=-7.14817522565231E+0000

Perihel
Abstand M-S = 3.07491007642847E-0001 AE
Geschwindigkeit= 5.89859100000002E+0004
Umlaufzeit= 8.79690137391701E+0001 Tage
Umlaufanzahl= 4.14999994494191E+0002   f= 4.15000000000000E+0002
Anzahl Durchgaenge= 3.15421691478465E+0010
Periheldrehung=-7.13552893862381E+0000

vo= 5.89859100000000E+0004
Zeit= 8.86966333333334E+0001 Minuten

The value of the mercury perihelion $\Delta\varphi=-7.136''$ done by the graviational acceleration g found above and which includes the natural mass rejection, is often reproduced, for example by a special view of planetary movement I call natural celestial Mechanics, without circular motion and angular momentum. It works only with movement speed and attack direction of the attraction. The trick is a radius of orbital motion that results from the components of the direction of attack. The components are the direction of movement and direction of fall at any time. This variant of celestial mechanics, however, requires a great performance of the computer. The rayon is calculated by $r=\dfrac{\left(s_u\right)^2}{2\cdot s_z}$. This is indeed confirmed by experiments, but not proven mathematically. s_z is the path perpendicular to the direction of movement within the shortest possible time, s_u is the way in the direction of motion in the same period.

The calculation above is done by steps of 0.1 seconds. This leads to a deviation of -0.026 seconds of arc. Based on the 50" of Orlov $\Delta\varphi_{Einstein}=\Delta\varphi_{Orlov}+\Delta\varphi$ the calculation gives a maximum value of 43.042 seconds of arc and a minimum value of 43.016" . Therefore the value of Orlov is full calculated by his $\Delta\varphi_{Orlov}=8.35\cdot10^{-8}\,rad$. The difference between 415 rounds of mercury and hundred years is considered.

It was found a relationship between the mass and the gravitational equivalent. As a reminder, it is

$$F=\frac{d}{dr}-\frac{1}{2}\Delta E$$

because $F=m\cdot a$. In the following, then using this relationship, the centrifugal force of a rotating body can be derived.
The time dilation of movement is $\tau=t\cdot\sqrt{1-v^2/c^2}$. If we remember the equation of c' by $c'=c\cdot\left(1-\dfrac{\Delta U}{c^2}\right)$ I must catch up on how I came to this relationship. The time dilation of gravitation is calculated by the equation $\tau=t\cdot\left(1-\dfrac{\Delta U}{c^2}\right)$. If our measurements of the speed of light do always give the same value back, it cannot be otherwise that the light goes the same way as the time dilation of gravitation. Our clocks and our hearts and everything is going with this time dilation. We are only able to measure a difference if we measure over a difference of height or doing this by counting steps in an airplane, but directly on the ground it is not possible to measure the time dilation. People in a fast starship will measure the same speed of light, too. Our second is the same second on every place and every movement. Also if the time is really dilated on our ground the speed of light has to be dilated in the same way. Additional there is to say, that this is a subjective fact only for the inhabitantt of the place, because he measures the same standards as everywhere. The standards are different over gravitation measured over a difference of height.

So it is the time dilation on which the above equation of force F resp. the relationship of the difference of mass equivalent and gravitation funds. By this understanding the equation of ΔE sounds otherwise

$$\Delta E=m\cdot c^2\cdot\left(1-\tau^2\right).$$

So that can be done with a potential having held body in movement, because movement has a time dilation, too. Rotation has a potential, but it is about reciprocally proportional to this of gravitation. By resuming the force law the acceleration a of a body in a potential sounds now

$$a=\frac{d}{dr}-\frac{1}{2}\cdot c^2\cdot\left(1-\tau^2\right)$$

So it is easy to derive the centrifugal acceleration of a rotating body. The time dilation τ has been $\tau=t\cdot\sqrt{1-v^2/c^2}$ and the rotation speed is $v_u=\omega\cdot r$. That leads to

$$a=\frac{d}{dr}-\frac{1}{2}\cdot c^2\cdot\left(1-\left(\sqrt{1-(\omega\cdot r)^2/c^2}\right)^2\right).$$

The centrifugal acceleration a results to

$$a=-\omega^2\cdot r$$

The minus results of the direction, centrifugal acceleration has towards gravitational acceleration.

In general, this phenomenon based on the indistinguishability between a stationary system in a gravitational field and an accelerated system, which has studied Paul Ehrenfest.

Cosmic Expansion Rate

In the universe governs the rejection. One speaks of accelerated expansion and suspects a Dark Energy which is responsible for it. The natural mass rejection is verified. Thus, these replace the dark energy.
Input was presented to the cosmological principle of this essay. The cosmos is like a balloon skin to be seen. Governed but in this balloon skin rejection, the balloon must be larger. The gravity has also been raised about the fourth dimension. To view the universe as a whole, it is primarily these, which need to be considered.

The cosmological conception was described. In this article, the four-dimensional gravity is similar to the spatial. We also know the speed specimens are able to reach, which fall to the ground. Among the mass rejection of these bodies would fly straight up the earth. Both can be described by the law of free fall of Isaac Newton and it was already used to derive the equation of gravitation, which involves the natural repulsion of mass, used in this paper. So it is easy to construct a relationship, but my experience told me the power of a Big Bang to make the construction work, because if there is only the free fall resp. free jump the values of the distance were not logical, are complex.

So, at first one note $\Delta E_{pot}=E_{kin}$. After that we remember $\Delta E_{pot}=\frac{1}{2}\Delta E$, so that one can

say $E_{kin}=\frac{1}{2}\Delta E$. As one know, the kinetic energy is $E_{kin}=m\cdot c^2\cdot\left(\dfrac{1}{\sqrt{1-\dfrac{v^2}{c^2}}}-1\right).$

It has been described, that Dark Matter is a result of the speed of enlargement of the balloon. The speed of enlargement is v in the equation and the mass m is the total baryonic mass of the universe - one has to notice, that the whole mass is in movement over the fourth dimension.

Will necessarily also be emphasized that the total moving mass at the same time the effective gravitational mass is, if you think about it. The mass moves itself, but not far enough. So additional there is the power of Big Bang E_{BB} necessary. If we now put all together it sounds:

$$m \cdot c^2 \cdot \left(\frac{1}{\sqrt{1 - \frac{v^2}{c^2}}} - 1 \right) = \frac{1}{2} \cdot \left(m \cdot c^2 - \frac{m}{\sqrt{1 - \frac{v^2}{c^2}}} \cdot \left(c \cdot \left(1 - \frac{m \cdot G}{\sqrt{1 - \frac{v^2}{c^2}} \cdot r \cdot c^2} \right) \right)^2 \right) + E_{BB}$$

For a better understanding reduced it sounds

$$(M - m) \cdot c^2 = \frac{1}{2} \cdot \left(m \cdot c^2 - M \cdot \left(c \cdot \left(1 - \frac{M \cdot G}{r \cdot c^2} \right) \right)^2 \right) + E_{BB}$$

which means

$$E_{kin} = \frac{1}{2} \Delta E + E_{BB}$$

The relationship is a little bit modified against our three-dimensional pendant. The rest mass m is in the first term of ΔE and the baryonic plus Dark Matter is in the second term. If we imagine the universe, at first it is in rest, so there is no Dark Matter. Dark Matter only arises with the expansion and the speed of it, the speed of the enlarging balloon. We remember the rule for subtraction in free fall: "Start minus target". This rule is used here, too. So Dark Matter plus baryonic is used in the second term for the gravitational potential and the mass in the start mass equivalent. The Big Bang power is added additional as basic energy.

In my experience it is best to take the middle equation and solve it to $M = f(r)$. So it is possible to pick better values of the steps i, what I want to say again before reading the full equation in the attachment. The equation varies over an r_n to r_u, where in n and u represents steps of i.

The problem now is to define r_n, which shall be the distance to the centre of the balloon in our present. Therefore we nearly know the relationship between m and M resp. ρ_{bary} and $\rho_{bary+dark}$. With M it is no problem to solve the middle equation to r_n, because the value represents values of the present. I take $\rho_{bary} = 4.5 \cdot 10^{-28} \, kg \cdot m^{-3}$ and $\rho_{bary+dark} = 3.0 \cdot 10^{-27} \, kg \cdot m^{-3}$ and create the relationship $q = \frac{\rho_{bary+dark}}{\rho_{bary}}$ and $Mq = m \cdot q$. So the relationship sounds

$$(Mq - m) \cdot c^2 = \frac{1}{2} \cdot \left(m \cdot c^2 - Mq \cdot \left(c \cdot \left(1 - \frac{Mq \cdot G}{r_n \cdot c^2} \right) \right)^2 \right) + E_{BB}$$

After this is solved to r_n. This gives two results:

$$R1 := \frac{1}{\left[2 \cdot \left(3 \cdot c^4 \cdot Mq - 3 \cdot m \cdot c^4 - 2 \cdot E \cdot c^2\right)\right]} \cdot \left(2 \cdot c^2 \cdot Mq^2 \cdot G - 2 \cdot \sqrt{-2 \cdot c^4 \cdot Mq^4 \cdot G^2 + 3 \cdot Mq^3 \cdot G^2 \cdot m \cdot c^4 + 2 \cdot Mq^3 \cdot G^2 \cdot E \cdot c^2}\right)$$

$$R2 := \frac{1}{\left[2 \cdot \left(3 \cdot c^4 \cdot Mq - 3 \cdot m \cdot c^4 - 2 \cdot E \cdot c^2\right)\right]} \cdot \left(2 \cdot c^2 \cdot Mq^2 \cdot G + 2 \cdot \sqrt{-2 \cdot c^4 \cdot Mq^4 \cdot G^2 + 3 \cdot Mq^3 \cdot G^2 \cdot m \cdot c^4 + 2 \cdot Mq^3 \cdot G^2 \cdot E \cdot c^2}\right)$$

The test shows after research, that R1 gives back a value that lays before the point of zero-g and R2 gives back the value of r_n behind zero-g. That is easy explained: If we have a look to M over r, we see

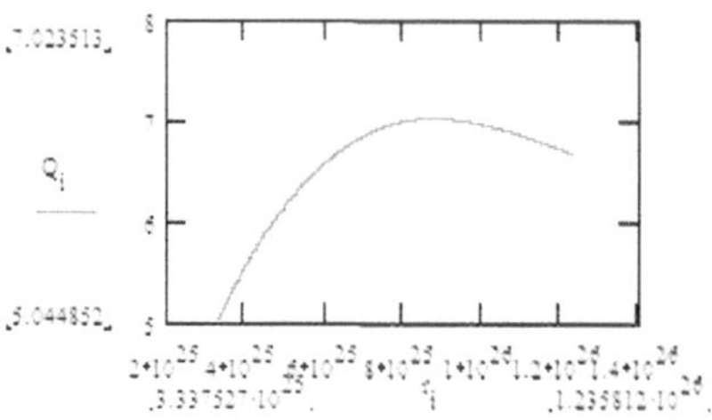

Q, wherever is the quotient q of baryonic plus Dark Matter to baryonic Matter. The graph shows two results for some Q's. These are R1 and R2. Because the rest mass has already been established in the beginning of the article, it must also fit the density of baryons matter, so that R2 only comes into question as a solution for our present. It is the bigger one. The point of zero-g is the top-maximum of the graph. After this the speed of enlargement decelerates, but only a few. This has to be shown down. The following results of $q \approx 5$. Rest mass and big-bang energy remained unchanged in the graphic shown down:

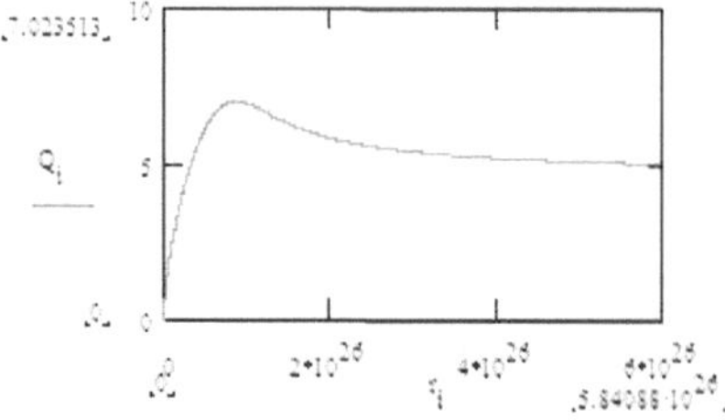

The equation of g shall be shown a second time specially to this:

$$g = \frac{M \cdot G}{r^2} - \frac{M^2 \cdot G^2}{r^3 \cdot c^2}$$

The equation cannot be solved for r, if one remember the long equation of M, but the point of zero-g can be found by the i-steps or by graphics. The graphic is shown down:

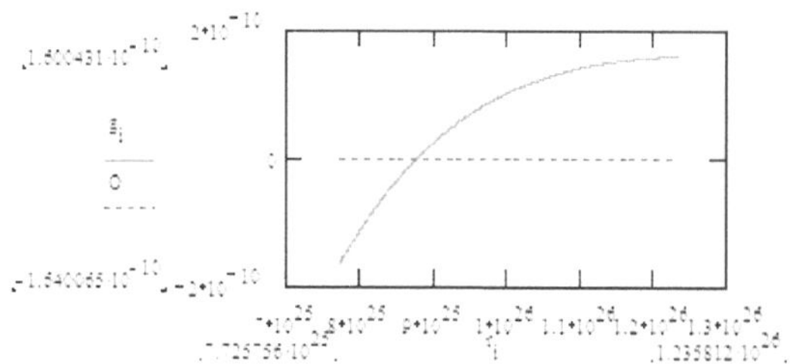

The blue curve in the chapter 'Results', which shows the theoretical distance r of a Black Hole is the same curve of solving

$$0 = \frac{M \cdot G}{r^2} - \frac{M^2 \cdot G^2}{r^3 \cdot c^2}$$

for r and it becomes the Event Horizont:

$$r_{g=0} = \frac{M \cdot G}{c^2}$$

As it is shown in the front, geometry and mechanics of the balloon defines the expansion rate, which is perceived in the balloon skin. So the speed of enlargement of the balloon v is necessary. This can be determined by m and M because M is known:

$$M = \frac{m}{\sqrt{1 - v^2/c^2}}$$

$$v = c \cdot \sqrt{1 - \frac{m^2}{M^2}}$$

Shows the speed v on r as follows:

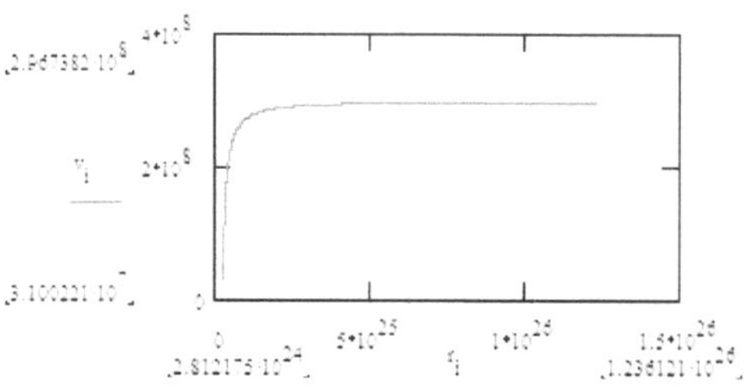

We realize, that the speed of enlargement v is near the speed of light most of the time. Because of the equation of v it was not possible to show the first value zero, but the dimension of the first value shows the great acceleration which exists in the first moment of Big Bang. One can call it an explosion.

It must be mentioned that the value for the big bang energy could only be found through trial and error. It amounts in this calculation $E_{BB}=8.34\cdot10^{69}\,Nm$. This is very, very much energy, much bigger than the mass equivalent of rest mass m. Maybe the cosm had a much bigger mass before the big bang, similar to a supernova, and most of it was converted into energy. The rest mass m also was found by trial and error but it came to the already shown value of $m=1.68\cdot10^{52}\,kg$ which leads to the value of $M=1.12\cdot10^{53}\,kg$ in the present, using the q which was told.

Now the cosmic expansion rate is to be found. So we have to have a look to the geometry of the perspective of an inhabitant in the skin of the balloon, which we call the universe and as it was begun at the beginning of the article.
At the beginning of the article there was shown an algorithm to find out the rest mass. It was not demonstrated mathematically, because it was not possible to close to r of the s. The s was demonstrated by a logical circuit. Now we have the r and it is not difficult to close to Δr of the time intervall t. It will be done by an average speed of enlargement of the balloon, because the full calculation is too complicated. While this the steps i helps:

$$t=\frac{\Delta r}{v}\text{ , what leads with }\Delta r=r_n-r_i\text{ and }v=\frac{v_n+v_i}{2}\text{ to }t=2\cdot\frac{r_n-r_i}{v_n+v_i}\text{ .}$$

That means, the way s will be calculated of the present base. The k, which was mentioned earlier, must also be based on an average speed v:

$$k=-\frac{v}{c}\text{ and with the average speed }k=-\frac{v_n+v_i}{2\cdot c}$$

We notice, k is negative because the logarithmic spiral must wind from present to past as our view. Now we can come to the angle φ. The basic equation of the logarithmic spiral

$$r=r_n\cdot e^{k\cdot\varphi}\text{ is solved to }\varphi=\frac{1}{k}\cdot\ln\left(\frac{r_i}{r_n}\right)\text{ resuming }v_e=\varphi\cdot v$$

As we know, the cosmic expansion rate H_0 *of the present* is defined by $H_0=\frac{v_e}{s}$ and $s=c\cdot t$. So H_0 can be shown down (O = 0):

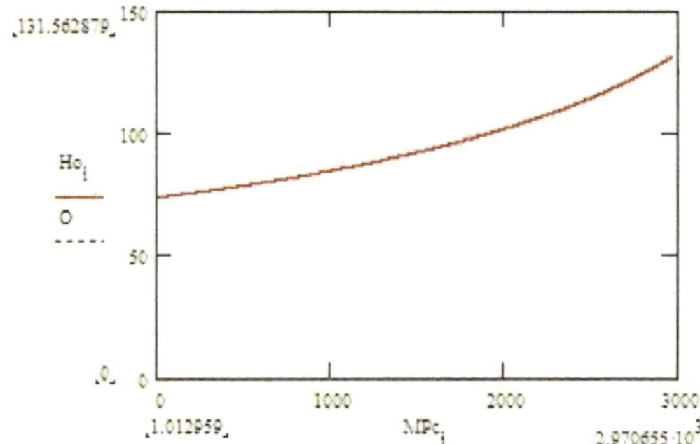

The graph to the black hole, which was once the universe, also shows that it was once smaller than a black hole. So the time would run around there wrong. The following chart clarifies this fact.

$$\tau = 1 - \frac{M \cdot G}{r \cdot c^2}$$

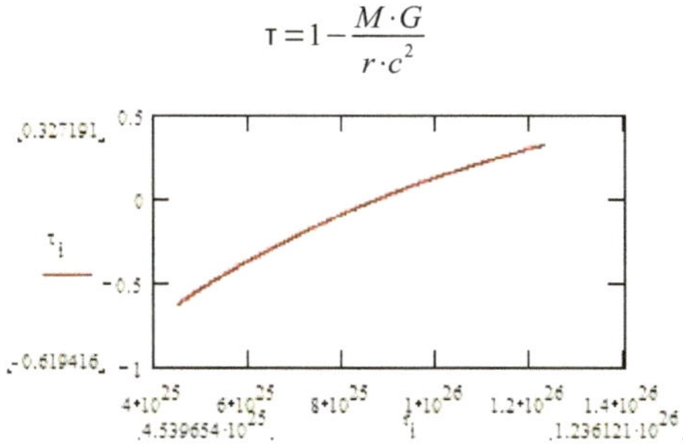

The problem must be seen from the perspective of the inhabitant of the balloon skin or the cosmos. To $v_e = c$ the difference of τ is $\Delta\tau = 0.947$, which means, the time course would not turn over, as far as one looks. $\Delta\tau = 1 - \left(\dfrac{\Delta\phi}{c^2}\right)$, $\Delta\phi = -\dfrac{M_n \cdot G}{r_n} - -\dfrac{M_i \cdot G}{r_i}$,

$$\Delta\phi = -\frac{M_{start} \cdot G}{r_{start}} - -\frac{M_{target} \cdot G}{r_{target}}$$

Now, there is to consider, that time dilatation reduces the power of light, the luminosity, if you look from far away in the time dilatated room. One can calculate this rediction by

$E_0 = E$, which means the dilatated energy is the same but the non-dilated energy, but with a bigger surface, a bigger sphere-radius S: $4 \cdot \pi \cdot s^2 \cdot 1 = 4 \cdot \pi \cdot S^2 \cdot 1 \cdot \Delta\tau$, so that $S = \dfrac{s}{\sqrt{\Delta\tau}}$. H

over S is shown down with S in Mpc by $H = \dfrac{v_e}{S}$.

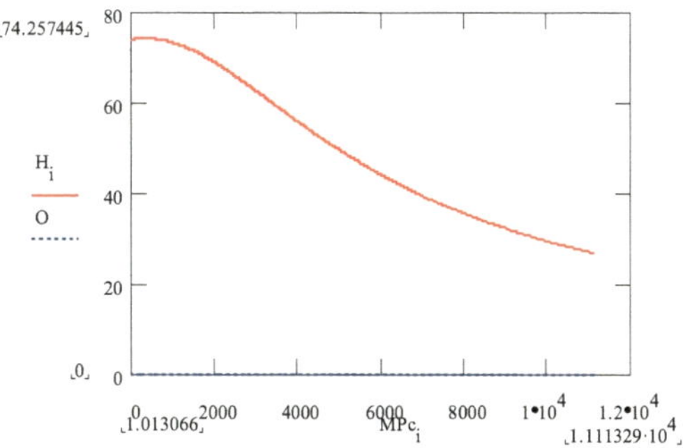

The expansion speed v_e over S looks

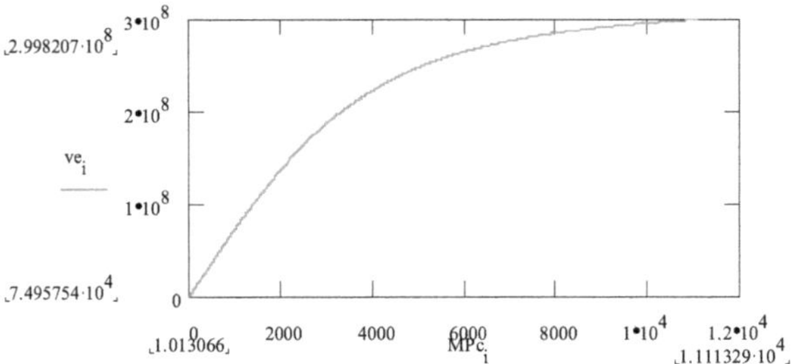

At least it is a nice feature to show the visible universe by the schematic representation, as if the universe were glassy, but reduced to two dimensions and further than the cosmic horizon allows:

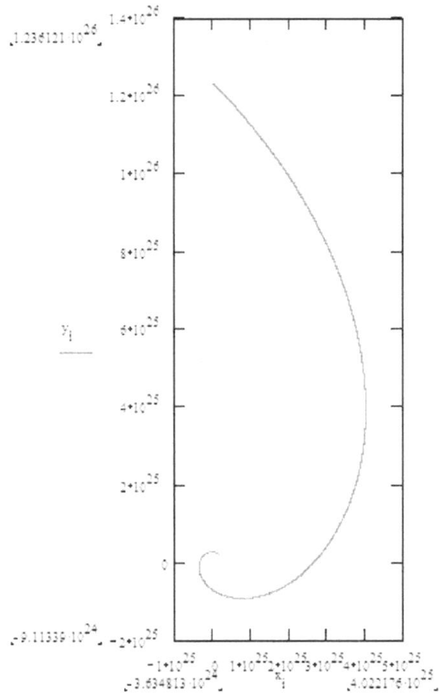

Not without interest is the fact of energy increase using the natural repulsion or antigravity. The following graphic is done by q=6.333. The other values remained unchanged.

The total energy of cosm is $\qquad$ $E_{tot} = M \cdot c^2$.

The originally existing energy $\qquad$ $E_{ex} = m \cdot c^2 + E_{BB}$.

The surplus energy $\qquad$ $E_{sp} = E_{tot} - E_{ex}$.

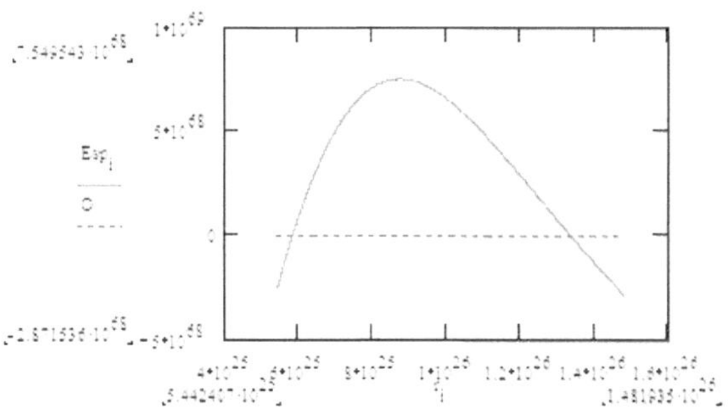

The graphic means that there is a surplus energy while the universe has a distance between $6 \cdot 10^{25}\, m$ and $1.35 \cdot 10^{26}\, m$ from the centre of its balloon. Thus, it is a temporary phenomenon, but it would be nice if it would be to use a technical. With the originally q it sounds like this:

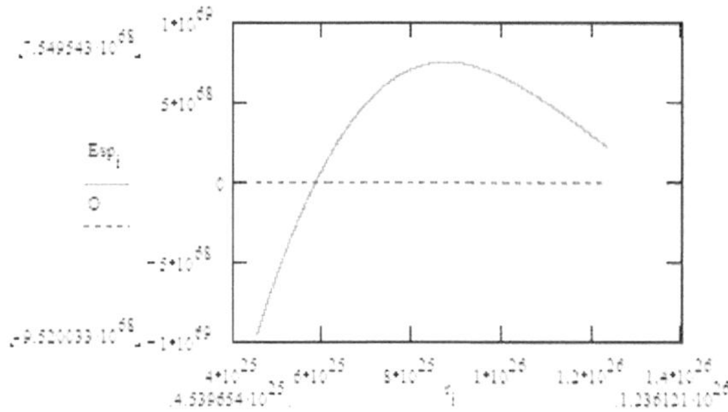

It can be assumed that the boundaries have not changed from the surplus energy. The study's is still pending.

Büdelsdorf, January the 5th, 2012

Attachment

1. Virtual Pascal v2.1 program to calculate the mercury-perihelion:

```pascal
program Periheldrehung; {Programm zur Ermittlung der Drehung des Perihels}
uses sysutils, crt;

var
                  z,k2,k3: integer ;
r,a,d,fi,Te,k,u,rv,vz,f,fif,Pd,vu,w,t1,d1,t3,v: extended;

const
AE=149.597870691e9; M=1.989e30; G=6.673848e-11; c=2.9979e8;
vo=5.898591e4; ro=46.00e9; rE=1*AE;                        t=0.1;

Begin t1:=time;

r:=ro; vu:=vo;
{d ist auf 415, der Rotationen in 100 Jahren, einzustellen}    d:=415;
{Venus:}

writeln('Merkur');
writeln(' ');

while fi<2*Pi*(d+0.1) do
Begin

w:=vu/r; fi:=fi+w*t;
a:=M*G/sqr(r)-sqr(M*G/(r*c))/r-sqr(w)*r;
vz:=vz+a*t; rv:=r; r:=rv-vz*t; vu:=vu*rv/r; {v:=sqrt(sqr(vu)+sqr(vz));}

Te:=Te+t{/(sqrt(1-sqr(v/c))-((-M*G/rE)-(-M*G/r))/sqr(c))};
k:=Te/t;

if fi>((d1-0.6)*2*Pi) then if r<rv then
begin
d1:=d1+1;
clrscr;
k3:=round(d1*78/(d+1));
for k2:=0 to k3 do write('X');
writeln(' ');
t3:=time;
writeln(' Noch ',round(24*60*((d*(t3-t1))/(fi/(2*Pi))-(t3-t1))),' Minuten')
End;

if fi>2*Pi*(d-0.6) then if z=0 then if r<rv then
begin
u:=fi/(2*Pi); f:=round(u*2)/2; fif:=f*2*Pi; Pd:=fi-fif; z:=z+1;
writeln(' ');
```

```pascal
writeln('Aphel');
writeln('Abstand M-S =',r/AE,' AE');
writeln('Geschwindigkeit=',vu);
writeln('Umlaufzeit=',Te/(3600*24*u),' Tage');
writeln('Umlaufanzahl=',u,'   f=',f);
writeln('Anzahl Durchgaenge=',k);
writeln('Apheldrehung=',Pd*648000/Pi);
writeln('   ');
End;


if fi>2*Pi*(d-0.15) then if z=1 then if r>rv then
begin
u:=fi/(2*Pi); f:=round(u*2)/2; fif:=f*2*Pi; Pd:=fi-fif; z:=z+1;
writeln('Perihel');
writeln('Abstand M-S =',r/AE,' AE');
writeln('Geschwindigkeit=',vu);
writeln('Umlaufzeit=',Te/(3600*24*u),' Tage');
writeln('Umlaufanzahl=',u,'   f=',f);
writeln('Anzahl Durchgaenge=',k);
writeln('Periheldrehung=',Pd*648000/Pi);
writeln(' ');
End;


End;


writeln('vo=',vo);
writeln('Zeit=',(time-t1)*24*60,' Minuten');
readln;
End.
```

2. Equation of M (total mass of the universe from baryons and Dark Matter):

(The equation is too long for presentation in this paper. Therefore, it is piecemeal but shown in the right order)

$$M_i := \left[\frac{-1}{54} \cdot c^2 \cdot \left(r_i\right)^2 \cdot \frac{\left(38 \cdot c^4 \cdot r_i - 81 \cdot G \cdot m \cdot c^2 - 54 \cdot G \cdot E\right)}{G^3} + \frac{1}{18} \cdot \sqrt{3} \cdot c^2 \cdot \frac{\left(r_i\right)^2}{G^3} \cdot \right.$$

$$\left. \sqrt{72 \cdot c^8 \cdot \left(r_i\right)^2 - 228 \cdot c^6 \cdot r_i \cdot G \cdot m - 152 \cdot c^4 \cdot r_i \cdot G \cdot E + 243 \cdot G^2 \cdot m^2 \cdot c^4 + 324 \cdot G^2 \cdot m \cdot c^2 \cdot E + 108 \cdot G^2 \cdot E^2} \right]^{\left(\frac{1}{3}\right)} - \frac{5}{9} \cdot c^4 \cdot$$

$$\frac{\left(r_i\right)^2}{\sqrt{G^2 \cdot \left[\frac{-1}{54} \cdot c^2 \cdot \left(r_i\right)^2 \cdot \frac{\left(38 \cdot c^4 \cdot r_i - 81 \cdot G \cdot m \cdot c^2 - 54 \cdot G \cdot E\right)}{G^3} + \frac{1}{18} \cdot \sqrt{3} \cdot c^2 \cdot \frac{\left(r_i\right)^2}{G^3} \cdot \sqrt{72 \cdot c^8 \cdot \left(r_i\right)^2 - 2} \right.}}$$

$$\frac{\left(r_i\right)^2}{\left. \left. \cdots^2 - 228 \cdot c^6 \cdot r_i \cdot G \cdot m - 152 \cdot c^4 \cdot r_i \cdot G \cdot E + 243 \cdot G^2 \cdot m^2 \cdot c^4 + 324 \cdot G^2 \cdot m \cdot c^2 \cdot E + 108 \cdot G^2 \cdot E^2 \right]^{\left(\frac{1}{3}\right)} \right]} + \frac{2}{(3 \cdot G)} \cdot c^2 \cdot r_i$$